SEVERE WEATHER

by Kathy Furgang

W9-DCN-119

TABLE OF CONTENTS

6
12
24
26

STORM Approaching!

Sometimes the weather has little effect on our day. Other times, it can change our lives forever . . .

"We just heard the news about Katrina on the radio, and I'm frightened more this time than with earlier hurricanes. Everyone in our town must evacuate. During other storms we were not ordered to leave our homes, but now we have to. Dad and Brian have covered the windows with planks, and Mom is frantically gathering food and clothes for all of us. We can bring only the bare necessities. Hurricane Katrina is just a day away and it is headed straight for us. We are not even sure where we are going. We'll just get in the car and drive north, away from the Gulf Coast. I hope our house is still here when we return."

▲ **Severe storms such as Hurricane Katrina can have devastating effects on people and property. Knowing how to stay safe can help people survive.**

Chapter 1

AIR in Motion

How does the movement of air cause severe storms?

When you got dressed this morning, did you stop to think about the weather? When you left home, did you take an umbrella or your sunglasses? Weather affects your daily life in many ways.

Weather is the condition of the atmosphere at any given time. Weather includes temperature, pressure, humidity, precipitation, cloud cover, and wind. Almost all weather happens in the **troposphere** (TROH-puh-sfeer). The troposphere is the layer of the atmosphere closest to Earth.

Weather is always changing. Weather changes from day to day, even moment to moment. One day might be cloudy and cool in the morning, and warm and rainy in the afternoon. Or one minute it's sunny and the next minute a storm blows in!

Hurricanes, tornadoes, and blizzards affect people around the globe every day. Severe storms are a weather phenomenon. What causes weather to change? What factors produce severe storms? In order to answer these questions, you need to know something about air in motion.

The Root of the Meaning

Troposphere

comes from the Greek words *tropo*, meaning "a change," and *sphaira*, meaning "ball."

Essential Vocabulary

anticyclone	page 8
cyclone	page 8
front	page 10
jet stream	page 10
troposphere	page 6

Weather is always changing—one minute it's sunny and the next minute a storm blows in!

Winds

At any given time, the sun warms some parts of Earth more than others. When air is warmed, it expands and becomes less dense. Less dense air rises and produces an area of low pressure. Cooler air contracts and becomes denser. Denser air is pulled closer to Earth and produces an area of high pressure. The cooler air moves in underneath the rising warm air. Air moves from an area of high pressure to an area of lower pressure and thus creates winds.

Strong winds are caused by large differences between areas of high and low pressure. Scientists call an area of low pressure that contains rising warm, circulating air a **cyclone** (SY-klone). In the Northern Hemisphere, the winds move around and into the center of a cyclone in a counterclockwise direction. The weather associated with a cyclone is usually rainy and stormy. In the Southern Hemisphere, these winds move in a clockwise direction.

Scientists call an area of high pressure that contains cold, dry air an **anticyclone** (an-tih-SY-klone). In the Northern Hemisphere, winds move around and out from the center of an anticyclone in a clockwise direction. The weather associated with an anticyclone is usually clear, dry, and fair. In the Southern Hemisphere, these winds move counterclockwise.

There are two general types of winds: local winds and global winds. Local winds blow from any direction and cover short distances. Most of the winds you are familiar with are local winds. Global winds blow from a specific direction and cover huge distances. They can affect areas over hundreds of square kilometers. Winds experienced before a big storm may be caused by global winds.

Global winds do not move in a straight line. Due to Earth's rotation, they blow around Earth's surface in a curved path. In the Northern Hemisphere, a wind curves to the right, and in the Southern Hemisphere, it curves to the left.

▲ **Weather associated with a cyclone is rainy and stormy.**

▲ **Weather associated with an anticyclone is clear, dry, and fair.**

SCIENCE TO SCIENCE:
PHYSICS AND METEOROLOGY

The Coriolis Effect

In 1835, French physicist Gaspard-Gustave Coriolis came to an important conclusion about forces on rotating objects. He determined that forces on a rotating surface move at right angles to the direction of the motion. This causes the forces to move in a curved line instead of a straight line. The Coriolis effect was an important concept in understanding and predicting the movement of winds that contribute to weather.

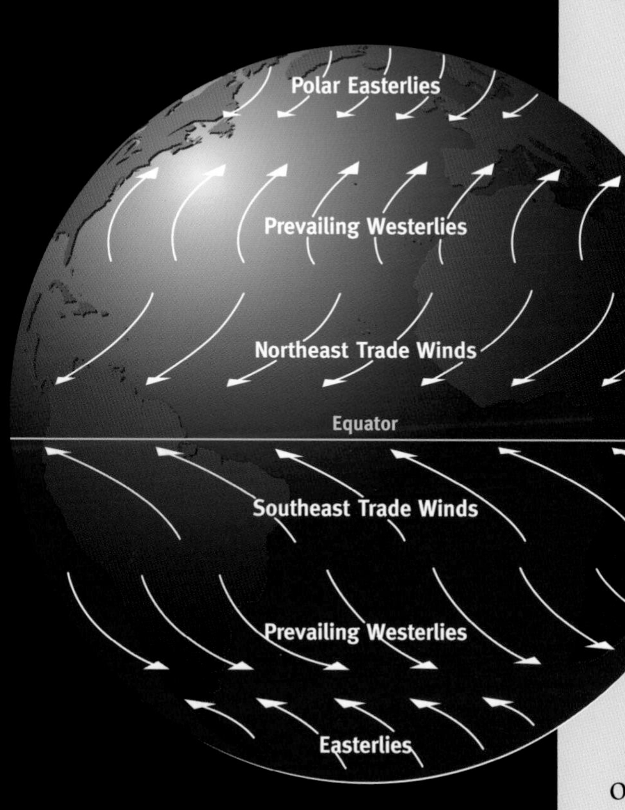

▲ The Coriolis effect is caused by Earth's rotation. This causes the curved paths of Earth's global winds.

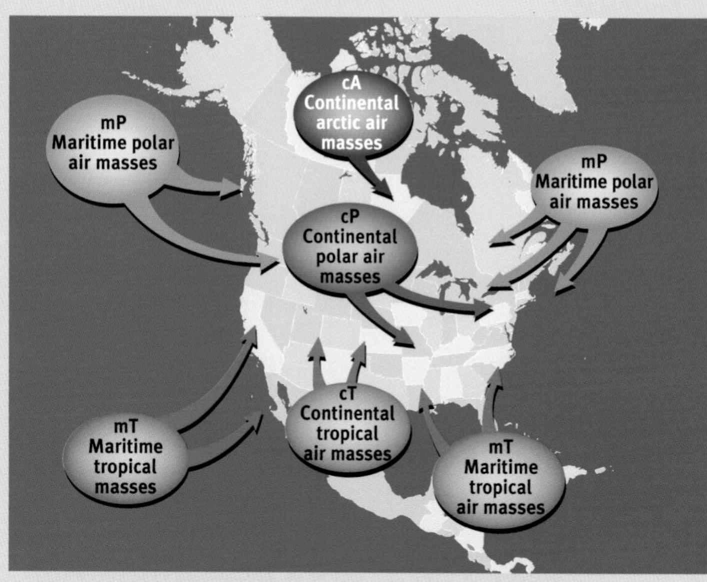

▲ There are six major types of air masses. Five of the six form over North America.

Air Masses

In addition to winds, large bodies of air called air masses are associated with changes in weather and severe storms. Air masses usually cover thousands of square kilometers. The properties of the air in an air mass—mainly temperature and moisture content—are nearly the same throughout the air mass.

Different air masses have different properties. Air masses are classified according to where they form. Air masses that form over tropical regions are warm. Air masses that form over polar regions are cold. Continental air masses form over continents and are relatively dry. Maritime air masses form over oceans and generally contain lots of moisture.

warm front

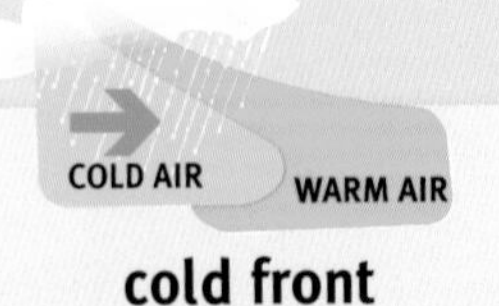

cold front

stationary front

occluded front

Fronts and Jet Streams

If you listen to weather reports, you have probably heard the word *front*. A **front** is a boundary that forms when two air masses with different properties meet. Most storms occur along this boundary. Look at the diagram above to see the four major fronts. Fronts travel across the globe according to the movement of jet streams. **Jet streams** are narrow belts of strong, fast-moving, high-pressure air found at altitudes of about 9 kilometers (5.6 miles) or more above Earth. Some jet streams move at up to 160 kilometers (100 miles) per hour. The minimum speed is 93 kph (58 mph). Some currents move as fast as 483 kph (300 mph).

Jet streams move in a curving motion from west to east around Earth, steering fronts and air masses. They become stronger if temperature differences between warm and cold air masses increase.

Powerful storms form when warm, light air rises quickly into higher, colder levels. Each type of storm forms under specific conditions. For example, hurricanes occur over oceans and coastlines because the air is rich with moisture. They draw their energy from warm ocean waters. Understanding the causes of severe storms allows us to predict when they may happen and also prepare for the damage and destruction they may cause.

SCIENCE AND TECHNOLOGY

A jet stream influences more than just the weather. It can influence air travel, too. In the United States, this high-speed wind system traveling from west to east is one of the reasons why it takes less time for a plane to travel from California to New York than it does to travel in the opposite direction.

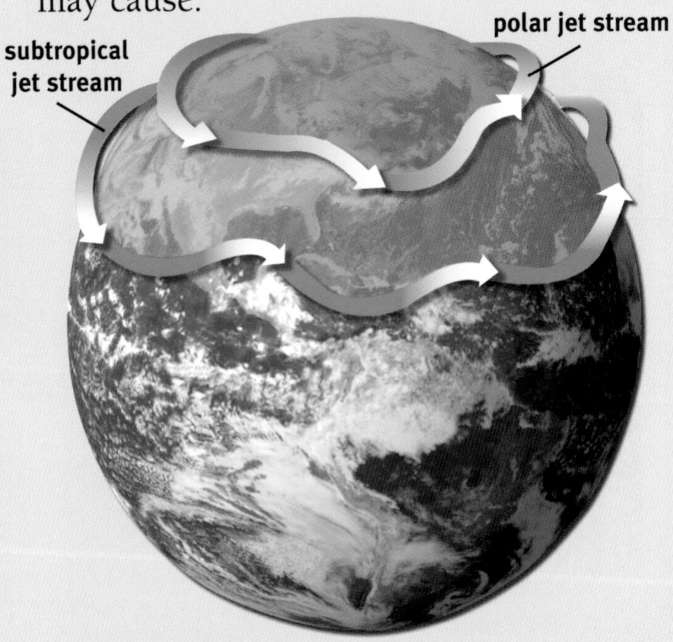

▲ Jet streams steer air masses around the globe.

SUMMING UP

- The unequal heating of Earth causes areas of low pressure and high pressure to develop.
- Areas of low pressure are called cyclones; areas of high pressure are called anticyclones.
- Air moves from high-pressure areas to low-pressure areas, creating winds. As a result of Earth's rotation, global winds blow around Earth's surface in a curved path.
- Air masses are large bodies of air with characteristic temperatures and moisture contents. When two air masses with different properties meet, a front forms. This is where most storms occur.
- Jet streams are narrow bands of high-speed winds that steer air masses and fronts from west to east around Earth.

PUTTING IT ALL TOGETHER

Choose one of the research activities below. Work independently, in pairs, or in a small group. Share your responses with the class.

1. Your weather forecaster reports that an anticyclone will be moving into your area. Reread page 8. Then write your own forecast in which you explain what an anticyclone is and describe what the resulting weather will be like.

2. Air masses are classified according to where they form. The five types of air masses over North America are continental polar, continental tropical, maritime polar, maritime tropical, and continental arctic. Use your school library or Internet resources to find out the characteristics of each air mass. Report your findings as a poster to share with the class.

3. Severe storms are usually associated with fronts. Find out what the four types of fronts are and what effects each front has on the weather. Summarize your findings in labeled diagrams.

Chapter 2

Types of SEVERE STORMS

WHAT ARE THE CAUSES AND CHARACTERISTICS OF VARIOUS SEVERE STORMS?

The destructive effects of a hurricane on a coastal town and the damage produced by a powerful tornado touching down in the Midwest may be familiar stories. So, too, may be the crippling effects of a blizzard. Have you experienced a thunderstorm? How about a blizzard or a hurricane? People all over the world experience these and other types of severe storms. Storms may be small and localized, or they may cover a huge area. They may form during a particular season of the year or in a specific geographic location. All storms are violent disturbances in the atmosphere that involve rising warm air, sudden changes in air pressure, and high winds.

▲ Lightning can occur between a cloud and the ground, between two clouds, between a cloud and the surrounding air, and within a cloud.

Essential Vocabulary

blizzard	page 16
El Niño	page 22
eye	page 18
flash flood	page 15
hurricane	page 18
La Niña	page 22
storm surge	page 19
tornado	page 20
waterspout	page 20

Chapter 2

Thunderstorms

A thunderstorm is a storm with strong rising air currents, thunder, lightning, and heavy rain or hail. Thunderstorms can produce winds of 93 kph (58 mph) or greater. At any given time, there are about 2,000 thunderstorms in progress around the world. Roughly 16 million thunderstorms occur each year. Most thunderstorms happen in the late afternoon after air has been heating all day. More thunderstorms occur in mountain regions where air rises and cools quickly.

Thunderstorms form when warm, humid air rises and cools rapidly. The moisture in the air condenses and forms a cloud. With further uplifting, the cloud extends higher, water droplets become larger, and some water droplets form ice crystals. At a height of about 10 to 20 kilometers (6 to 12 miles), a cumulonimbus (KYOO-myoo-loh-NIM-bis) cloud, or thunderhead, forms.

Thunderstorms are often preceded or accompanied by heavy rain or hail. When water droplets in a cumulonimbus cloud get large and heavy, they fall to the ground as rain. If the droplets become ice crystals and the crystals are uplifted over and over again, they become so heavy that they fall to the ground as hailstones.

Lightning and Thunder

The quick movement of air inside a cumulonimbus cloud causes the water droplets and ice crystals to become electrically charged. Positive and negative electrical charges build up in the cloud and move to opposite ends. When a large difference between the charges builds up—either within the cloud, or between the cloud and the ground—a discharge of electrical energy occurs. This discharge is a lightning strike that appears as the brilliant bolt of light you see. Thunder is the sound made when lightning quickly

Flash Floods

When a large amount of rain falls in a very short period of time, a **flash flood** can occur. Flash floods are most common in areas close to mountains. When heavy rain falls faster than the ground can absorb it, water begins to collect on the ground. Mountain streams rise higher than the stream banks, and the streams overflow in a matter of hours. Water flows downhill, adding to the water already collecting at the bottom of the mountain. Flash floods can also happen in dry areas. Rain falls so quickly that the ground cannot absorb it.

Flash floods can be dangerous. Water levels may rise 9 meters (30 feet) or more. The moving water can be strong enough to sweep cars along with it. Water carrying heavy debris can crush bridges and trees, and sweep away objects in its path.

EVERYDAY SCIENCE

Electrical Conductivity and Safety

Because lightning is an electrical discharge, the rules of electrical conductivity apply to being out in stormy weather. A conductor is a material through which electricity flows easily. Water and many metals are considered conductors. So outdoor water sources, such as pools, rivers, or oceans, should be avoided. So, too, should contact with metals and telephone wires. The best insulators in a storm are things made of rubber, glass, or wood.

▼ Flash floods can carry away objects such as cars, trees, and street signs.

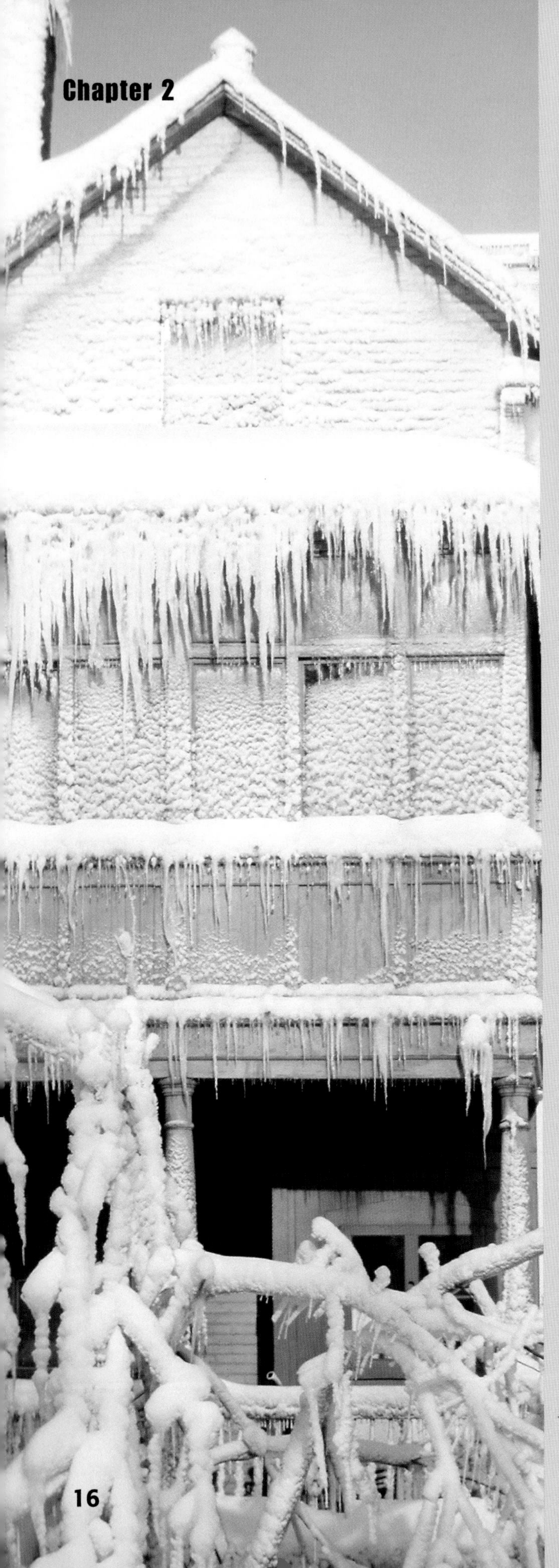

Winter Storms

Snow can be fun to play in, and tree limbs encased in ice can be pretty to view. But make no mistake, winter storms can be dangerous and destructive. Winter storms are weather events in which cold temperatures produce precipitation in the form of snow, sleet, or freezing rain. In temperate regions, these storms are usually associated with winter. However, they can occur in late autumn or early spring.

Winter storms usually form when two different air masses meet. The northeastern United States experiences many winter storms due to its latitude. Cold, dry, polar air from the north combines with warm, tropical air from the south. As the warm air rises and cools, water vapor in the clouds condenses. As the droplets of condensation get large and heavy enough, they fall from the clouds as precipitation. Snow is formed in below-freezing air. Below-freezing temperatures between the clouds and the ground keep the snow from melting before it hits the ground.

A **blizzard** is a snowstorm with below-freezing temperatures, winds of 56 kilometers (35 miles) per hour or more, and visibility that is reduced to less than 0.4 kilometer (0.25 mile). Blizzards are caused by warm air rising over colder air. These storms can happen quickly, which makes them very dangerous. People can become trapped by the storm. Strong winds and heavy snow can also cause

◀ **The weight of ice during an ice storm can damage trees and other property. Power outages can cause difficulties and dangers for local residents.**

power outages. Snowdrifts can block roads. Snowdrifts are banks of deep snow that form when wind blows snow to form piles that are more than double the amount of snowfall in depth.

▲ **Blizzards are caused by warm air rising over much colder air.**

One of the worst blizzards occurred in 1888. The "Great White Hurricane," as it was called, lasted two days without stopping and blanketed the East Coast from Virginia to Maine in more than 120 centimeters (48 inches) of snow. Snowdrifts up to 15 meters (50 feet) high were recorded.

Ice storms are another dangerous type of winter storm. Rain that turns to ice in the process of falling can make surfaces slick and hazardous. Heavy ice coats everything. This coating can cause branches to break off and fall, damaging telephone and power lines.

Today, scientists can predict upcoming blizzards and other winter storms. The most accurate tool scientists use is Doppler radar. It measures the direction and speed of objects in the air, such as individual snowflakes, raindrops, or ice crystals. The radar indicates exactly where a storm is, where it is headed, and how quickly it is traveling.

SCIENCE TO SCIENCE: LIFE SCIENCE

What happens to animals during storms? Not all animals survive sudden bouts of severe weather, but a surprising number of them do. Animals with dens or burrows in the ground can escape rain and dangerous lightning until the storm passes. Others can sense when a storm is approaching. They may feel a change in air pressure in their ears, they may see the decrease in light due to dark clouds, or they may be sensitive to increasing winds. This gives them enough time to seek shelter.

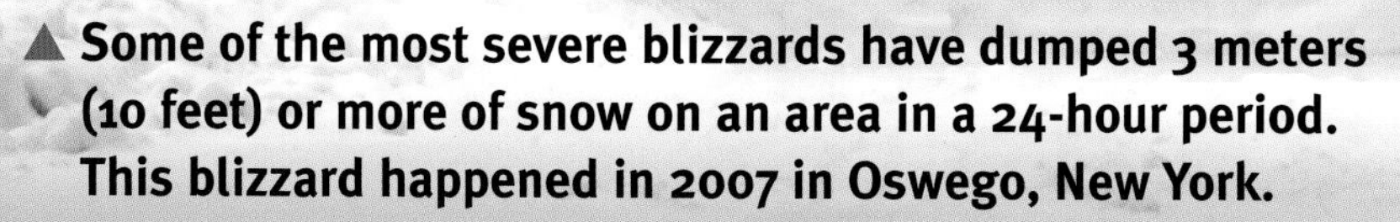

▲ **Some of the most severe blizzards have dumped 3 meters (10 feet) or more of snow on an area in a 24-hour period. This blizzard happened in 2007 in Oswego, New York.**

▲ Hurricanes are the most violent storms on Earth.

Hurricanes

Hurricanes are powerful cyclones (low-pressure areas) that form over warm, tropical oceans near the equator. Here warm, moist air rises rapidly and small thunderstorms form. Clusters of these thunderstorms gather and begin to rotate counterclockwise around a common low-pressure center. One huge spiraling storm with high winds results. When the wind speeds are between 37 and 63 kilometers (23 to 39 miles) per hour, it is called a tropical depression. When the wind speed reaches 63 kilometers (39 miles) per hour, it is called a tropical storm. When the wind speed reaches 119 kilometers (74 miles) per hour or more, the storm is categorized as a hurricane.

Hurricanes are usually 400 to 800 kilometers (250–500 miles) in diameter, so they affect a large area when they make landfall. The low-pressure center of a hurricane is called the **eye**. The weather in the eye is surprisingly calm. However, just outside the eye is the eye wall. This is where the strongest winds blow and the heaviest rain falls. The eye wall is responsible for the greatest damage a hurricane causes. Some of the worst hurricanes have been known to kill or injure hundreds of thousands of people and devastate towns.

Hurricanes are the most violent storms on Earth in terms of their total energy. A typical hurricane lasts for about nine days, but some have lasted as long as three to four weeks. Because a hurricane is fueled by warm ocean water, it begins to lose its strength and eventually break up when it hits land.

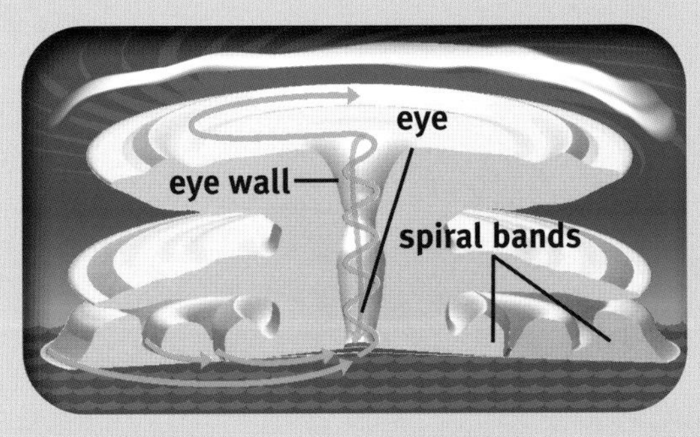

▲ Tropical storms with winds greater than 119 kph (74 mph) are called hurricanes.

Storm Surge

In addition to the damage caused by hurricane winds, a great deal of damage results from a hurricane's **storm surge**. As a hurricane moves across the ocean, its winds push large amounts of ocean water forward in front of the storm. As the hurricane reaches land, so too does this wall of water, which is called a storm surge. The storm surge can cause the sea level to rise by as much as 6 meters (20 feet).

The size of a storm surge depends on the speed and strength of the hurricane, and whether it reaches land during high or low tide. Storm surges can cause extensive damage to property on the shoreline, both before and as the hurricane makes landfall.

Checkpoint
Visualize It

Choose one of the storms described in this chapter and draw a picture that shows the effects of this type of storm.

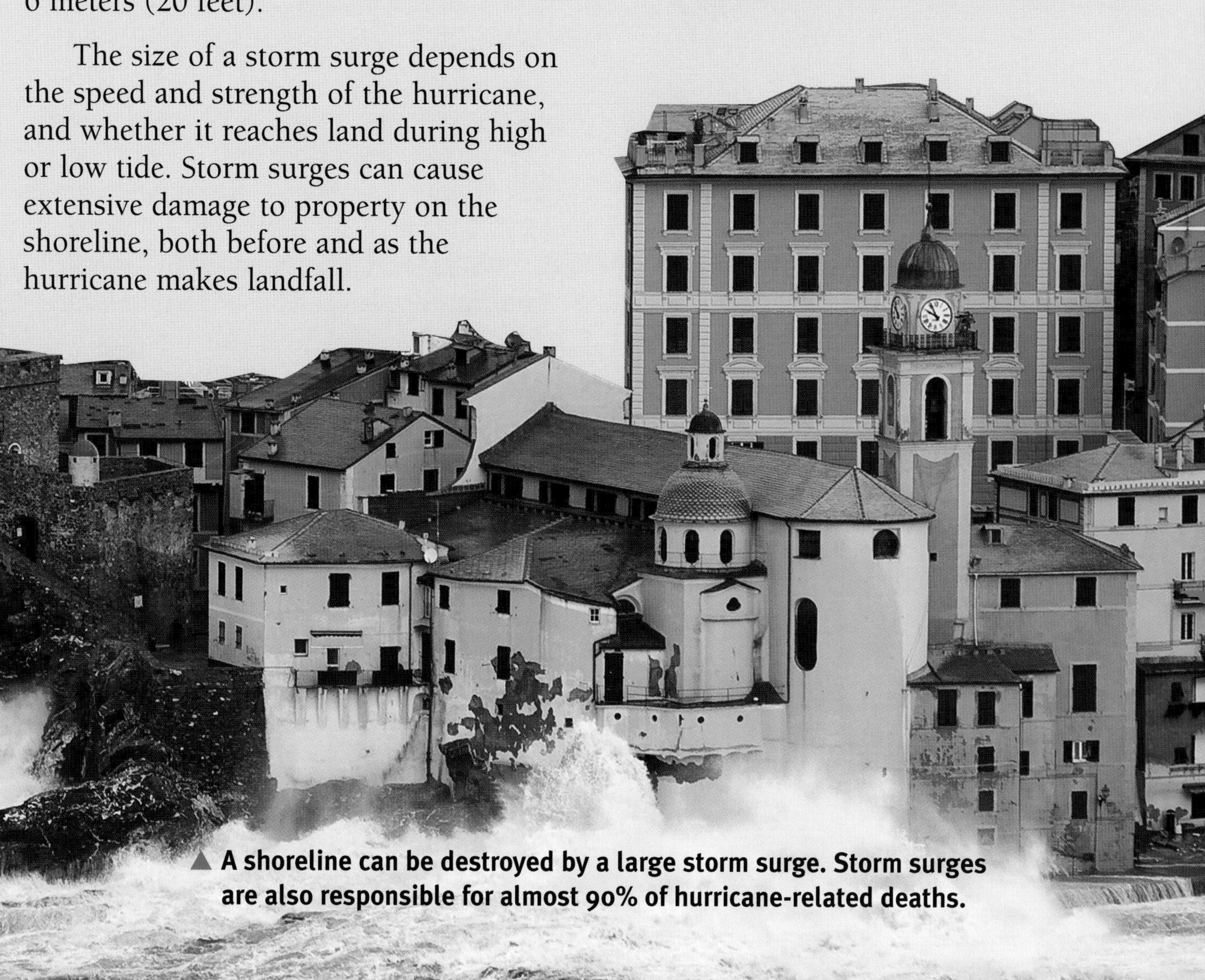

▲ A shoreline can be destroyed by a large storm surge. Storm surges are also responsible for almost 90% of hurricane-related deaths.

Tornadoes

A **tornado** is a powerful, swirling funnel of rising air that comes in contact with both the clouds and the ground. A tornado is often called a twister because of its spinning, twisting motion. Some small tornadoes have wind speeds around 64 kilometers (40 miles) per hour. The largest, most destructive tornadoes can be a kilometer wide with wind speeds up to 483 kilometers (300 miles) per hour.

Tornadoes can cause massive destruction where they touch the ground, leveling buildings and sweeping up everything in their paths. But, because the part of the tornado that touches the ground is small, the path of destruction is often very narrow. Sometimes the tornado even skips over an area. So the houses on one street might be destroyed, while the houses on the next street are undamaged.

▼ Most tornadoes last only a few minutes. The dark, funnel-shaped cloud contains dirt and debris swept up by the tornado.

Tornadoes form during extremely strong thunderstorms. They can also form in the aftermath of a hurricane. Varying currents of air cause the thunderstorm to spin. When winds inside the storm move at different speeds, or in different directions, a column of air inside the storm begins to spin. If this spinning column of air is fed by warm, moist air from below, the winds can begin to spin even faster and a tornado can form.

A **waterspout** is a tornado over water. Waterspouts are usually weaker than tornadoes over land, but they can still be strong enough to suck objects out of the water and carry them to other locations. The objects are dropped when the waterspout weakens. Waterspouts have been known to carry fish, frogs, or other objects considerable distances from where they were picked up.

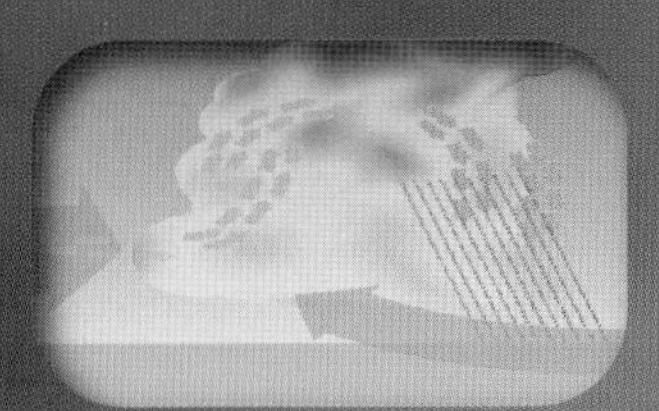

1. Strong thunderstorms create an updraft.

2. High winds create a spinning column of air.

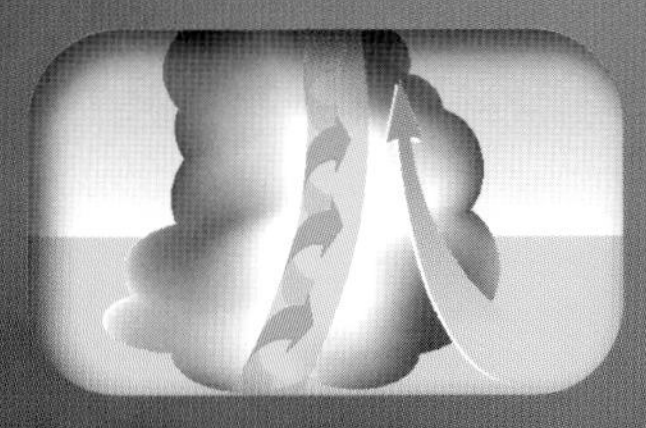

3. The updraft tilts the spinning column up toward the sky.

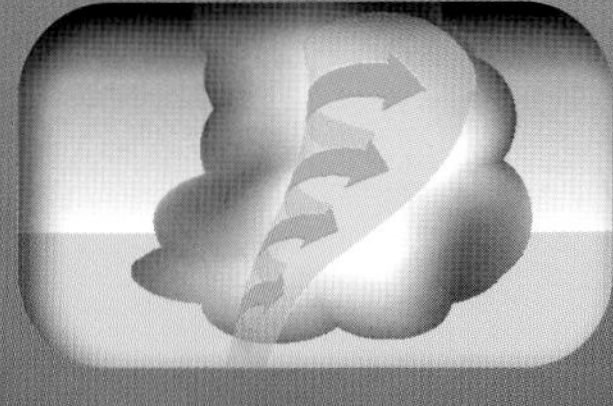

4. Winds grow stronger as warm air rushes in from below.

Hands-On Science

A tornado is a whirling, funnel-shaped cloud. How can water be made to move in a funnel shape to illustrate a tornado cloud?

Time

20 minutes

Materials

2 empty plastic 2-liter soft drink bottles, steel or rubber washer to fit one of the bottle openings, duct tape, water

Procedure

1. Fill one bottle 2/3 full with water, and tape the washer to the top of the bottle.

2. Tape the second bottle to the top of the first bottle so that their openings are touching. Use as much tape as necessary to make the seal as secure and watertight as possible.

3. Turn the bottles over so the water is now in the top bottle. Quickly swirl the bottles several times. Place them on a table and observe.

Analysis

What happens to the flow of water from the top bottle to the bottom bottle?

El Niño and La Niña

Changes in the atmosphere's wind patterns can cause major weather problems around the world. **El Niño** (EL NEEN-nyoh) is a disruption in the wind patterns and water currents in the tropical Pacific Ocean. El Niño causes severe weather disturbances around the globe.

Usually, winds blow strongly from east to west along the equator, pushing the warm water toward the western part of the Pacific Ocean. In the eastern part of the Pacific, deeper, colder water along the coast of South America gets pulled up from below to replace the water pushed west. This upwelling is important because it brings nutrient-rich water to the surface. These nutrients support large fish populations.

About once every three to four years, an El Niño forms. The trade winds along the eastern Pacific Ocean weaken and the warm ocean water that piled up in the west returns to the eastern Pacific Ocean. Less cold water gets pulled up from below. The water in the eastern Pacific gets much warmer than usual. This change affects weather around the globe. Heavy rain and serious flooding can occur in the eastern Pacific Ocean. At the same time, severe drought (DROWT) occurs in the western Pacific.

The term **La Niña** (LAH NEEN-nyah) is used to describe the event in which the Pacific Ocean waters along the coast of South America are much cooler than normal. La Niña also affects climate and weather around the globe. The effects of La Niña are the opposite of El Niño.

El Niño and La Niña may cause extreme rains and flooding in some parts of the world or severe droughts in other parts.

Checkpoint
Read More About It

Use library or Internet resources to find out more about El Niño. How does it affect the fishing industry, coral reefs, birds, and marine animals?

SUMMING UP

- Thunderstorms, blizzards, hurricanes, and tornadoes are examples of severe storms.
- Although each storm has its own characteristics, all are violent disturbances in the atmosphere that involve rising warm air, sudden changes in air pressure, and high winds.
- El Niño and La Niña are regularly occurring changes in wind and water currents that cause severe storms and drought around the globe.

PUTTING IT ALL TOGETHER

Choose one of the research activities below. Work independently, in pairs, or in a small group. Share your responses with the class.

1. Reread page 22, and then do further research about El Niño and La Niña. Create a collage that summarizes the effects of these weather patterns. What different types of severe weather can occur around the globe?

2. Use resources in your school or public library or on the Internet to find out about some of the most destructive hurricanes of the past twenty years. Where did they originate? Where did they make landfall? What was the extent of the damage? Write a short report to summarize your research.

3. Reread page 14 about thunderstorms. Create a diagram to illustrate the formation of this storm and its accompanying lightning. Be sure to label the diagram appropriately.

TORNADO ALLEY

CARTOONIST'S NOTEBOOK • BY PETE PACHOUMIS

WHAT ARE SOME WAYS WE CAN TELL A STORM IS COMING? WHAT WOULD YOU DO IF A TORNADO, HURRICANE, OR BLIZZARD WERE HEADED YOUR WAY? DO YOU HAVE AN EMERGENCY PLAN?

Chapter 3

Storm WARNINGS AND SAFETY

HOW CAN PEOPLE REMAIN SAFE DURING SEVERE STORMS?

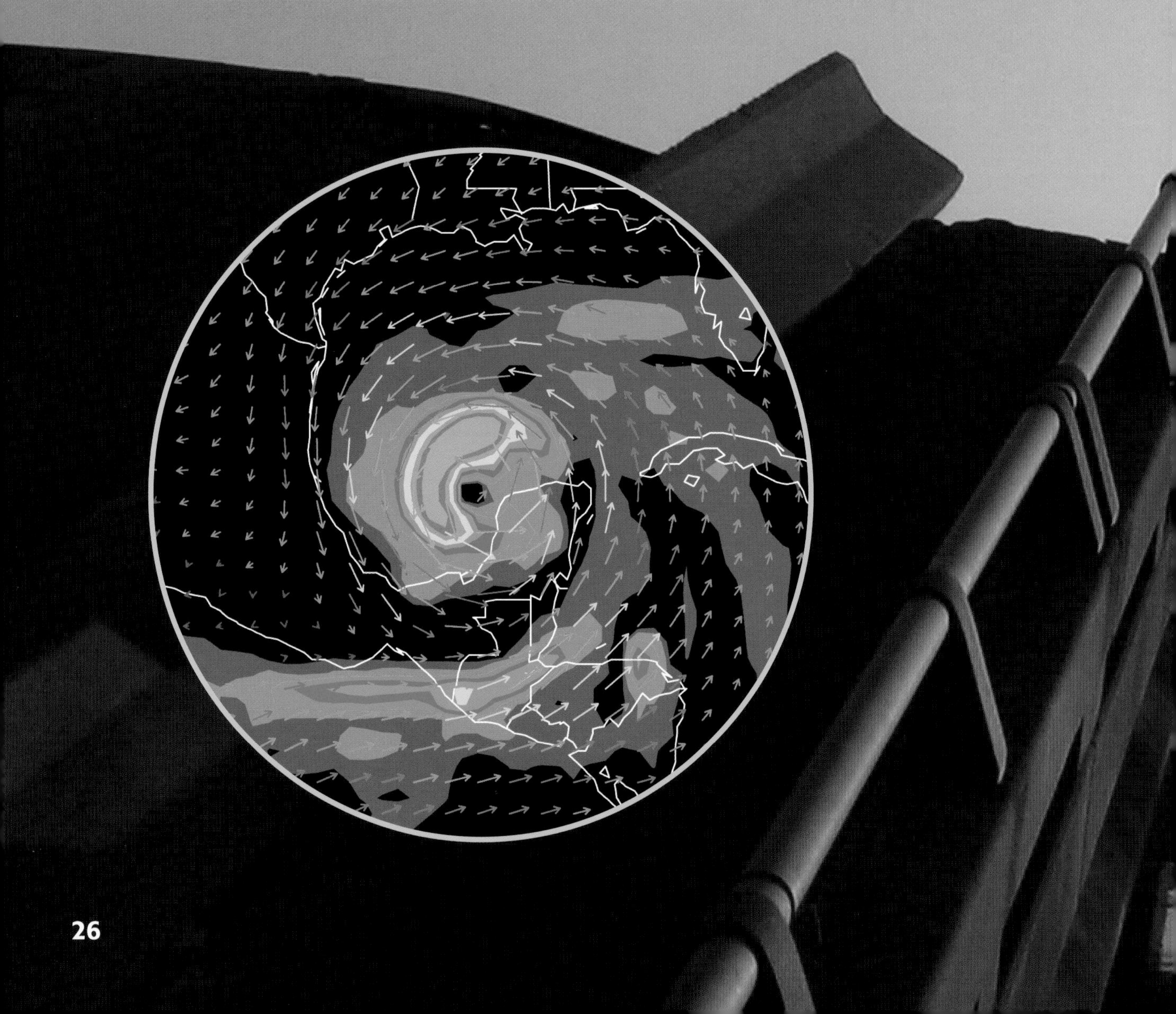

Essential Vocabulary

evacuation	page 38
Fujita Scale	page 33
Saffir-Simpson Hurricane Scale	page 30

Each year in the United States, thousands of people are injured or killed by severe storms. These storms also cause billions of dollars in property damage. In 2006, for example, tornadoes killed 67 people, injured 990 people, and damaged over 75 billion dollars of property. In that same year, floods from storms killed 76 people, injured 23 people, and caused over 37 billion dollars of property damage.

In 2005, more than 1,000 people died as Hurricane Katrina devastated the Gulf Coast region. The storm also caused property damage amounting to over 93 billion dollars.

Fortunately, the total number of people killed or injured each year from severe weather in the United States has decreased. This is a result of more accurate weather predictions and better warnings.

Hurricane Katrina hit the Gulf Coast in 2005 and was one of the deadliest hurricanes in the United States. It was a Category 3 storm.

National Weather Service

The National Weather Service (formerly known as the U.S. Weather Bureau) is the federal agency responsible for forecasting and warning the public about any storms or severe weather events. The National Weather Service has offices across the United States. Meteorologists at each office track and predict the weather in their local area. They analyze weather data gathered from radar and satellites. They keep detailed records of weather conditions.

The National Weather Service alerts the public about severe weather, such as blizzards, hurricanes, and thunderstorms. A weather advisory, watch, or warning will appear on television, radio, and the Internet. This lets people prepare for bad weather. Local authorities follow the recommendations of the National Weather Service. They may close schools, airports, or other public facilities.

How the National Weather Service tracks storms and warns the public depends on the type of storm. There are different ways for people to prepare for these events.

Predicting Hurricanes

The Atlantic hurricane season runs from June 1 to November 1. Beginning in June, meteorologists at the National Weather Service start to keep a close eye on the ocean waters near the equator. That is where the hurricanes that hit the Atlantic and Gulf coasts form. Using satellite photos and other tools, meteorologists can spot tropical depressions, which are the clusters of thunderstorms that develop into tropical storms and then hurricanes.

The meteorologists carefully track storms, which usually develop in the eastern Atlantic Ocean and move toward the western Atlantic Ocean. They gather data and analyze satellite photos so they know when a tropical depression has turned into a tropical storm, and when it has developed into a hurricane. During an average hurricane season, the National Weather Service will track approximately sixteen tropical storms. About eight of these storms will become hurricanes.

SCIENCE TO SCIENCE: DOPPLER RADAR

Doppler radar uses a principle called the Doppler effect, named after Austrian physicist Christian Doppler, who lived during the first half of the 1800s. He noticed an apparent change in the frequency of sound waves depending on the location of the observer. The closer the observer is to the source, the greater the frequency. For example, have you ever heard an ambulance siren as it drives past you? As the ambulance approaches, the siren is very loud. The moment it passes, however, the sound gets lower and is not as loud as it moves away. The moving ambulance is actually pushing the sound forward, which makes the sound's frequency higher in the direction it is moving. Frequency readings can help identify the location, direction, and speed of a storm.

Meteorologists are scientists ▶ who study the weather.

In the past, large numbers of people died in severe storms, because they often hit with no warning. Today, when a hurricane appears to be heading toward land, the National Weather Service will issue hurricane watches and warnings to the public. A hurricane watch means that hurricane conditions are possible within thirty-six hours. A hurricane warning means that hurricane conditions are expected within twenty-four hours.

National Weather Service meteorologists use hurricane warnings and watches to forecast the storm's path. They also inform people about the strength of the storm using the **Saffir-Simpson Hurricane Scale**. The scale assigns a number to each hurricane according to its wind speed.

Forecasts from the National Weather Service are critical because they help government officials and people living in the area prepare for the storm. Usually when a hurricane hits, people on the coast need to evacuate, or leave their homes.

Leaving a Hurricane Zone

People in the path of some hurricanes need to evacuate because the conditions during the storm are so dangerous. The winds are powerful enough to blow out windows and tear roofs off houses. For places on the coast, the storm surge brings a wall of water sweeping across the land. The rushing water is powerful enough to pick up and carry boulders, cars, and other large objects. Conditions after the storm are dangerous as well. Large areas of land can be flooded. Often there is no electricity or running water. It can take time before it is safe for people to return to their homes.

Government officials use the National Weather Service's forecasts to help them decide who must evacuate. If meteorologists are predicting a Category 1 storm, officials might tell people living directly on the water and people living in mobile homes to evacuate. If a Category 3 or 4 hurricane is predicted, then officials will order a much larger evacuation.

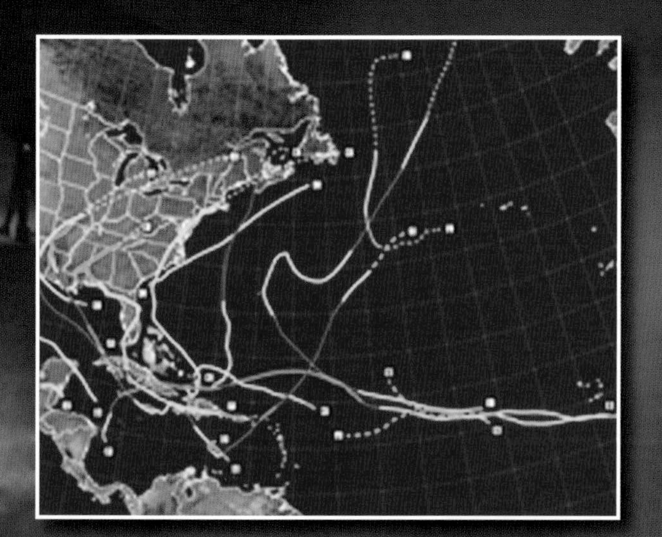

Experience has shown that using people's names is faster and more accurate than using coordinates when tracking storms. There are ten tropical storm regions worldwide. Each region has a rotation of six alphabetical lists of names that they use each year. Deadly and costly hurricanes, typhoons, or cyclones are usually retired from the list. You can see the lists for each region at www.nhc.noaa.gov.

Saffir-Simpson Hurricane Scale

Hurricane Scale	Wind Speeds	Possible Damage
Category 1	119–153 kph (74–95 mph)	Damage to trees, shrubs, unanchored mobile homes
Category 2	154–177 kph (96–110 mph)	Extensive damage to trees, shrubs, mobile homes; increased flooding
Category 3	178–209 kph (111–130 mph)	Increased property damage; requires evacuation of people living in low-lying coastal areas
Category 4	210–250 kph (131–155 mph)	Complete destruction of some properties; flooding 10 kilometers (6 miles) inland
Category 5	Greater than 250 kph (155 mph)	Complete failure of most roofs; some homes swept away; total evacuation of people up to 16 kilometers (10 miles) from the ocean

Fujita Scale

▼ This scale measures the strength of a tornado. It ranks tornadoes from F0 to F5 based on the damage the tornado causes.

Category	Wind Speed	Damage
F0	116 kph (72 mph)	Light—Some damage to chimneys; branches broken off trees; shallow-rooted trees pushed over; sign boards damaged
F1	117–180 kph (73–112 mph)	Moderate—Peels surface off roofs; mobile homes pushed off foundations or overturned; moving cars pushed off the roads; attached garages may be destroyed
F2	181–253 kph (113–157 mph)	Considerable—Roofs torn off house frames; mobile homes demolished; boxcars overturned; large trees snapped or uprooted; light-object missiles generated
F3	254–332 kph (158–206 mph)	Severe—Roofs and some walls torn off well-constructed houses; trains overturned; most trees in forest uprooted; heavy cars lifted off the ground and thrown
F4	333–418 kph (207–260 mph)	Devastating—Well-constructed houses leveled; structures with weak foundations blown away some distance; cars thrown and large missiles generated
F5	419–513 kph (261–319 mph)	Total—Strong frame houses lifted off foundations and carried considerable distances to disintegrate; automobile-size missiles fly through the air in excess of 100 meters (109 yards); trees debarked; steel-reinforced concrete structures badly damaged; incredible phenomena will occur

Predicting Tornadoes

The National Weather Service also forecasts tornadoes and issues tornado watches and warnings. But unlike hurricanes, which travel slowly, tornadoes form quickly and with much less warning. That, and the fact that scientists don't completely understand how and why tornadoes form, makes forecasting them difficult.

National Weather Service meteorologists look for the conditions that can lead to tornadoes. Certain times of the year and locations in the United States make tornadoes more likely. For example, tornadoes typically

CAREERS IN SCIENCE: STORM CHASERS

A storm chaser is a person who pursues storms in order to gather data, make observations, or photograph events. Many storm chasers work independently and sell their work to news channels or magazines. They often begin by checking weather radars and then driving directly toward the storms to be the first to detect any changes. They often record the storm as an eyewitness account.

form during the spring in the Midwest when warm moist air from the Gulf of Mexico meets cold, dry air from the North. This produces the severe thunderstorms that can generate tornadoes.

National Weather Service meteorologists use data about air temperature and wind patterns to determine if thunderstorm conditions exist. Radar is helpful in detecting the formation of thunderstorms. When tornadoes seem possible, a tornado watch is issued. When a tornado is spotted touching down, a tornado warning is issued. Meteorologists use wind speed and reported damage to rate the tornado on the **Fujita Scale** (see page 32).

Tornado Safety

People living in the Midwest have to pay close attention to forecasts during tornado season. If a tornado watch is issued, they need to be prepared to seek shelter. If a tornado warning is issued, they must seek shelter immediately. Finding shelter in a basement, under a staircase, or under a heavy table offers protection from falling debris or crumbling walls. Interior rooms that have no windows, such as bathrooms, closets, or hallways, are good places to seek shelter. Standing in a doorframe is another option. Glass from shattered windows flies in a tornado and can cause injury, so it is important to stay away from windows.

Tornadoes approach quickly, which means there is little time for people to prepare. People living in tornado zones need to be alert to the signs that a tornado is near in case they do not hear a tornado warning on the news.

▲ **With a tornado approaching, finding a place that is protected from flying debris is important.**

TORNADO WARNING SIGNS

- **The sky takes on a greenish or greenish-black color.**
- **Clouds move very quickly, especially in a swirling pattern in one area of the sky.**
- **There is a loud sound of rushing air, like a waterfall, jet, or train.**
- **Debris falls from the sky.**
- **A funnel-shaped cloud forms in the sky.**

The yellow shaded area on the map represents Tornado Alley, the region of the United States with the highest concentration of tornadoes.

Chapter 3

Winter Storm Safety

During the winter, the National Weather Service issues advisories for severe winter weather. A winter storm watch means a possibility of a storm in the area. A winter storm warning means a storm is headed that way.

▼ **Blizzards can cause whiteout conditions in which visibility is reduced to almost nothing.**

The National Weather Service also issues blizzard warnings. Blizzards are the most dangerous types of winter storms. To stay safe, people should remain indoors during a blizzard. Except in an emergency, people should avoid traveling. Drivers can be stranded when blowing snow and snowdrifts make roadways impassable. Drivers should keep emergency supplies, such as extra blankets, water, and food, in the car.

Safety Guidelines for Severe Weather

Whatever type of severe weather occurs, the National Weather Service offers general suggestions for staying safe.

Families should have an emergency kit prepared. The kit should contain gallons of bottled drinking water and nonperishable food. There should be enough food and water to last three days. The kit should also include a flashlight, extra batteries, a first-aid kit, extra blankets, and a battery-powered radio. Power outages can happen during all types of severe weather. Families will need the radio to get up-to-date weather information if there is no electricity.

Immediately before a storm, people should listen closely to weather reports and directions from local authorities. In some cases, people will be required to evacuate, or leave, the area.

SCIENCE TO SCIENCE: CHEMISTRY & ENGINEERING

Engineers are always searching for new ways to make buildings stronger, especially in tornado-prone areas. Wind speeds up to 402 kilometers (250 miles) per hour in the center of a tornado can send objects into buildings with the strength of a torpedo. Scientists are working to develop stronger concrete by combining limestone and clay at very high temperatures. New types of concrete might be able to withstand more wind damage, and that just might be enough to save lives.

An emergency kit should be prepared ahead of time to include essential items in case people have no heat, air conditioning, or electricity.

Chapter 3

Evacuation Plans

In the case of an **evacuation**, people must move to a new place until the dangers of a storm have passed. The Federal Emergency Management Agency, or FEMA, recommends that families have an evacuation plan. Think ahead of time about where your family might go in case of an evacuation. The destination should be clearly out of the way of the storm and should not take too long to reach. Choose a contact person whom everyone in your family can call in case you get separated. If the possibility of an evacuation exists, the family car should be filled with gas and a battery-powered radio should be kept in it. Listen to the radio for instructions, and then follow the instructions carefully.

Checkpoint
Talk It Over

What do you do to get ready for a storm? Do you live in an area of the country that gets a lot of tornadoes, hurricanes, floods, or winter storms? Talk in a group about how your family prepares for dangerous storms.

Sometimes a boat is the only way to get to safety during a flood.

After the Storm

The danger does not disappear immediately after the storm has passed. It is important for people to exercise caution after the severe weather. Dangerous conditions, such as downed power lines and hazardous debris, may exist. The local water may not be safe to drink. Roads may not be passable.

Hurricanes and thunderstorms can cause flooding. Walking through a flooded area, especially if the ground cannot be seen through the water, is hazardous. Flowing water can knock people off their feet. There is also the danger of electrocution. Even after people have returned to their homes, it is important that they follow the instructions of local officials.

▲ An evacuation can take a day or more to complete, especially in populated areas.

CAREERS IN SCIENCE: DISASTER RELIEF WORKER

Severe weather takes a toll on communities. People may be left homeless or injured. There may be extensive property damage, or people may be stuck inside homes or vehicles. Trained doctors, nurses, and construction workers are among the relief workers who volunteer their time at disaster sites and help communities recover from trauma.

	Type of Storms	Characteristics	Safety Precautions
Hurricane		• most violent storms on Earth • winds at 119 kph (74 mph) or more • heavy rain • flooding • form over warm tropical waters	• evacuate to shelter that is clearly out of path of storm • keep emergency supplies such as fresh water, food • be aware of downed power lines, trees, flood currents, falling debris
Blizzard		• heavy snow • winds at 56 kph (35 mph) or more • whiteout • form when warm air rises over colder air at below-freezing temperatures	• stay indoors • avoid traveling unless in emergencies • keep emergency supplies such as blankets, food, water • be aware of icy, slippery streets, downed power lines, sliding snow on roofs
Tornado		• destructive funnel of severe wind • winds at 64–513 kph (40–319 mph) • form from severe thunderstorms	• seek shelter in a basement, under a staircase, or under a heavy table • choose interior rooms without windows such as closets, halls, bathrooms • be aware of downed power lines, falling debris, broken glass, faulty structures
Thunderstorm		• heavy rain • winds at 93 kph (58 mph) or greater • lightning and thunder • hail up to 2 cm (3/4 in) • tend to form in late afternoon or over mountains	• stay indoors • stay away from trees, water, open fields • avoid appliances • if you are in a car, remain inside • be aware of downed power lines, trees, flooding

SUMMING UP

- The National Weather Service issues forecasts, watches, and warnings to keep people informed about approaching storms and the possible need to evacuate.
- Each type of storm has characteristic warning signals.
- General guidelines for storm safety always apply: keep an emergency kit handy; listen closely to weather reports and directions from local authorities; prepare an evacuation plan; and remain alert and cautious after the storm has passed.
- Severe storms can devastate a community, but lives can be saved if people know how to prepare for storms and follow safety procedures.

PUTTING IT ALL TOGETHER

Choose one of the research activities below. Work independently, in pairs, or in a small group. Share your responses with the class.

1. Think about what you and your family might do if you had to evacuate your home because of a severe storm. Where would you go? What route would you take to get there? What physical features of your town (nearness to mountains or water) do you need to consider? Write a step-by-step evacuation plan for your family. Present it to the class and compare your plan with those of your classmates.

2. Use library resources or the Internet to find out more about the United States National Weather Service. What kind of weather instruments do they use? How do they collect information? How do they communicate this information to the public? Present your findings in a visual report.

3. Look at the map of Tornado Alley on page 35. Use that map and library or Internet resources to identify the states included in Tornado Alley. Then find out what other areas of the United States have the greatest occurrence of tornadoes. Identify the conditions that make tornadoes so common to these areas. Finally, find out where else in the world tornadoes occur frequently. Present your findings to the class.

STORM Aftermath

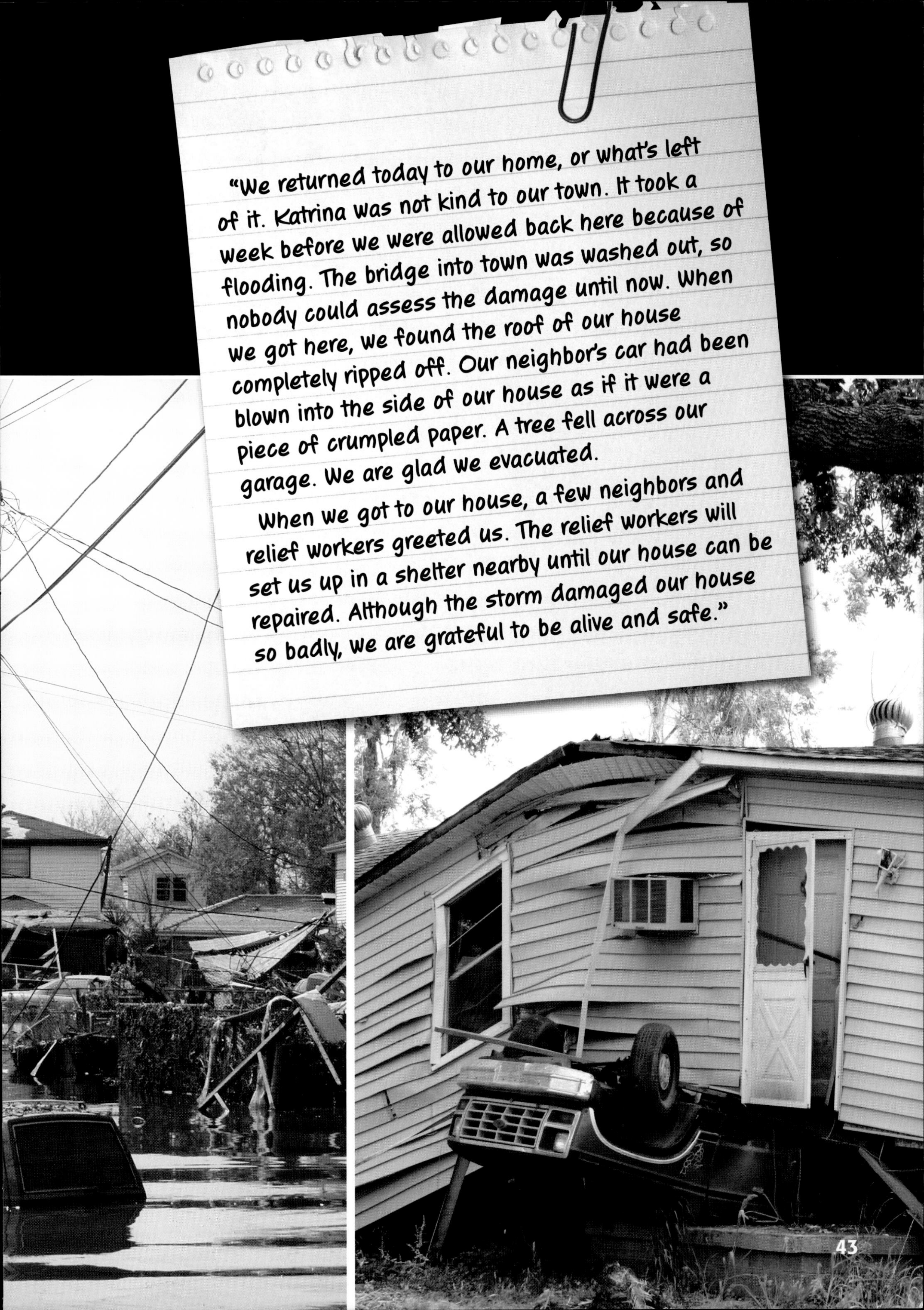

"We returned today to our home, or what's left of it. Katrina was not kind to our town. It took a week before we were allowed back here because of flooding. The bridge into town was washed out, so nobody could assess the damage until now. When we got here, we found the roof of our house completely ripped off. Our neighbor's car had been blown into the side of our house as if it were a piece of crumpled paper. A tree fell across our garage. We are glad we evacuated.

When we got to our house, a few neighbors and relief workers greeted us. The relief workers will set us up in a shelter nearby until our house can be repaired. Although the storm damaged our house so badly, we are grateful to be alive and safe."

How to Write an Account

Not all of people's everyday experiences may be worth recounting, or retelling for scientific purposes. However, if someone has had a unique experience, recording the details of that experience may be helpful to scientists investigating common characteristics of a particular event, such as a storm.

A scientific account should include as many details as possible about the event. If you are recounting a storm experience, here are a few guidelines for you to follow.

Record the physical details of your environment, not your emotional reactions to the storm. Take note of any changes in temperature, cloud patterns, wind patterns, or water levels in rivers, oceans, or streams. Scientists can compare these details to other accounts of similar storms. Note the differences in the environment before, during, and after a storm. Was there precipitation? If so, at what time did it begin? If flooding occurred, how quickly did it happen? How high was the water level? At what time did the storm end? What type of damage did you observe? Never risk your safety to record observations. Your safety and the safety of those around you should be your first priority. You can always record your observations once the storm is over.

▲ Scientists can learn more about storms through people's accounts of the experience.

For example, people who have experienced and recounted a tornado have identified common experiences. Many recall the greenish-black sky color before the storm hit, the fast-moving clouds that take on a swirling pattern, and finally the tell-tale funnel-shaped cloud in the sky. People also tell similar tales of hearing loud, rushing air making the sound of a jet or train. Scientists have heard accounts of frogs or fish falling from the sky. Such information has helped scientists better understand the nature of waterspouts.

At 4 P.M., the sky turned a greenish-black color. Then the fast-moving clouds began to swirl in the sky. Then the tornado came through fast and furious. It sounded like a freight train. And just as quickly as it came, it was gone. We lingered for a few minutes longer in our designated safe zone and then emerged. Outside, the wind was calm and the air was quiet. We looked across the street and not a thing was out of place. You would never know a storm had been here. Then we looked at the Raymonds' house to the left of us. It was completely gone, missing from its foundation. We could see directly into their basement, except that the big old oak tree from our yard lay across it. None of us had ever seen anything like this. We could see the direct path where the tornado had gone through the street. We were lucky. We lost only our oak tree and some shutters and shingles from the house. But our neighbors have to start all over.

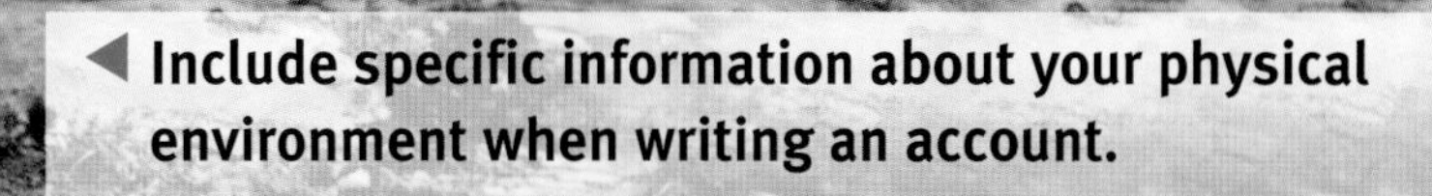

Include specific information about your physical environment when writing an account.

Glossary

anticyclone	(an-tih-SY-klone) *noun* high atmospheric pressure system moving in a clockwise direction in the Northern Hemisphere (page 8)
blizzard	(BLIH-zerd) *noun* storm with below-freezing temperatures, winds of 56 kilometers (35 miles) per hour or more, and reduced visibility (page 16)
cyclone	(SY-klone) *noun* low atmospheric pressure system moving in a counterclockwise direction in the Northern Hemisphere (page 8)
El Niño	(EL NEEN-nyoh) *noun* an unusually warm current in the Pacific Ocean that moves eastward along the equator from Asia to the South American coast (page 22)
evacuation	(ih-va-kyoo-AY-shun) *noun* when an entire population leaves an area due to unsafe conditions such as severe winds, flooding, or precipitation (page 38)
eye	(I) *noun* the low-pressure center of a storm (page 18)
flash flood	(FLASH FLUD) *noun* when large amounts of rain fall in very short periods of time and cannot be soaked into the ground (page 15)
front	(FRUNT) *noun* an area where a cold air mass meets a warm air mass (page 10)
Fujita Scale	(foo-JEE-tuh SKALE) *noun* a rating system that assigns numbers to tornadoes according to their wind speed and strength (page 33)

hurricane (HER-ih-kane) *noun* storm forming in the ocean with winds that reach 119 kilometers (73 miles) per hour or more (page 18)

jet stream (JET STREEM) *noun* fast-moving air about 9 kilometers (5.6 miles) above Earth (page 10)

La Niña (LAH NEEN-nyah) *noun* unusually cool surface temperatures in the eastern Pacific Ocean along the South American coast (page 22)

Saffir-Simpson Hurricane Scale (SA-feer-SIMP-sun HER-ih-kane SKALE) *noun* a rating system that assigns numbers to hurricanes according to their wind speed and strength (page 30)

storm surge (STORM SERJ) *noun* rising of sea level that occurs as a result of wind and storm systems (page 19)

tornado (tor-NAY-doh) *noun* a powerful, swirling funnel of rising air that comes in contact with both the clouds and the ground (page 20)

troposphere (TROH-puh-sfeer) *noun* the closest atmospheric layer to Earth (page 6)

waterspout (WAU-ter-spowt) *noun* tornado that forms over water (page 20)

Index